Anonymous

Pulmonary Phthisis and the Value of Hypophosphites of Lime and Soda in Its Treatment

Anonymous

Pulmonary Phthisis and the Value of Hypophosphites of Lime and Soda in Its Treatment

ISBN/EAN: 9783337817022

Printed in Europe, USA, Canada, Australia, Japan

Cover: Foto ©berggeist007 / pixelio.de

Can Consumption be Cured?

We appreciate the endorsement and generous support that the Physicians of the United States and British Provinces have given to DR. McARTHUR'S CHEMICALLY PURE SYRUP OF HYPOPHOSPHITES. Our special claim to your favor is the CHEMICAL PURITY of our SYRUP, and its adaptability as a base for the addition of the salts of Iron, Quinia and Strychnia, as may be indicated in special cases, but not always desirable in the Syrup.

Boston, Mass.

October 17th, 1886.

PULMONARY PHTHISIS,

AND THE VALUE OF

HYPOPHOSPHITES OF LIME AND SODA

IN ITS TREATMENT.

DR. CHURCHILL popularized the hypophosphites in the treatment of pulmonary phthisis more than a quarter of a century ago. The benefit which has since attended their use has been such as to give them a highly meritorious place amongst the therapeutic agents that have been employed in the treatment of this disease.

Pulmonary consumption produces a greater ratio of mortality than any disease to which the human body is liable, and any remedy, therefore, which favors recovery should receive the highest consideration.

The morbid changes in the lungs of patients with pulmonary phthisis are caused by *tubercles*. According to the germ theory, the disease is caused primarily by the *tubercle bacillus* discovered by Koch. Exactly how this microbe does its work has not yet been determined. All investigators, however, grant that *environment* exerts a powerful influence, not only in protecting the individual from all germ diseases, but in aiding in overcoming them after they have gained an entrance into the human body. For example, tuberculosis may become active only in consequence of the presence of favorable conditions, either in the person or in his surroundings. Hence the great importance of observing the precautions to be adopted in the treatment of consumption laid down on page 46 of the pamphlet "Can Consumption be Cured?"

All those agencies which reduce the vital, resisting power of the tissues favor the development of pulmonary tuberculosis. This includes exposure to air vitiated by any cause, and the non-observance of well-established hygienic and sanitary regulations. The *greatest* resistance is offered to the development of phthisis by those who have an active circulation, unimpaired digestion, and a normal condition of all the physiological functions. In such persons nutrition is maintained to its highest degree, which is the strongest safeguard against the invasion of any disease. True, there are individual differences in resisting power; as, for instance, one individual may escape an attack, though repeatedly exposed to a contagious disease; but these cases are exceptions, and strengthen the rule that such exposure is followed by the development of the affection.

Cases of pulmonary phthisis can be classed under four heads, and their treatment, at the present time, is both hygienic and medicinal. The first includes those cases in which the disease is arrested; the second those in which its progress is slow; the third those in which recovery takes place; and the fourth those in which the disease progresses steadily towards a fatal termination. The last class includes, by far, the largest number. It is a fair assumption, however, that the favorable course pursued in the cases belonging to the three other groups has been the result of *judicious treatment.* Although some patients tolerate phthisis much better than others, the imperative fact remains that favorable terminations are seen almost exclusively in those cases which have been under treatment. The opinion is unanimous that treatment must be begun *early* to be the most effectual. The hacking cough, the slight fever, the least emaciation, etc., should prompt the immediate use of the *Hypophosphites.* Their general tonic properties, the aid which they afford to digestion, and the tone and energy which they impart to the nervous system, are all required to place the patient beyond the destructive processes belonging to tubercles.

But the existence of cavities in the lungs does not banish the hope of obtaining slow progress by treatment. The disease may also be arrested after excavations have formed in the lungs, and the patient may enjoy a fair degree of health for many years. Some cases end in recovery even after cavities of considerable size have been formed. Pulmonary phthisis manifests a natural tendency to recovery in only a small number of cases. The good results, therefore, which have been reported have been obtained by hygienic and judicious medicinal treatment. To this end the *Hypophosphites* have been used extensively. Precisely how they act may be questioned, but that they exert a decidedly beneficial influence is the opinion of eminent clinical observers. It must, then, be admitted, in the absence of a specific cure for phthisis, "that it is the imperative duty of the physician," said the late Prof. Flint, "to bring to bear upon it any agency, medicinal or hygienic, which experience shows to have a favorable influence, however small." This eminent authority reported cases in which the patients *recovered* while taking the *Hypophosphites*. These, together with those reported by other observers, still sustain the great popularity of the remedy.

AITKEN quotes COGHILL, who, as the result of large hospital experience, regards "the hypophosphites as one of the best general tonics in phthisis."

BARTHOLOW believes that "the hypophosphites are valuable agents to promote constructive metamorphosis."

LOOMIS says that "when intestinal indigestion is imperfect the hypophosphites are especially beneficial."

PALMER writes that "the hypophosphites have obtained a noticeable reputation" in the treatment of phthisis, and that "by their use a needed ingredient is added to the system, and nutrition is improved."

JOHN TAYLOR, as the result of his experience in the use of the hypophosphites, in private practice and in the Liverpool Workhouse and Infirmary, says that, "in the earlier and the middle stages of phthisis they produce a glowing influence as a respiratory excitant, expanding the chest, increase the appetite and cheerfulness, and control expectoration, night-sweat and diarrhœa." He believes that their influence should be guarded by a tonic, and his favorite remedy is the compound tincture of gentian.

RICHARD PAYNE COTTON, physician to the Brompton Hospital for Consumption, London, says that, "in cases associated with acid dyspepsia and requiring alkaline treatment, the hypophosphites of lime and soda are often successful."

JAMES R. BENNETT of London says that "it is extremely probable that the hypophosphites will prove useful in the treatment of phthisis, acting beneficially on the stomach and intestinal mucous membrane, and, as Dr. Churchill affirms, on the pulmonary mucous membrane, diminishing expectoration and giving tone to the capillaries."

J. C. THOROWGOOD, M. D., physician to the Hospital for Diseases of the Chest, Victoria Park, London, Eng., reports the following cases as illustrative of the beneficial effects produced by the hypophosphites:

"J. R. E., a pale, thin young man, dated his illness from a sudden spitting of blood. The left side of his chest was flattened, with impaired percussion resonance, and abundant crepitant rale in inspiration. Cod-liver oil had always made him sick. On the previous day he brought up blood. He was ordered to take hypophosphite of soda in camphor water three times a day. May 17th, seven days afterward, the medicine agreed well, and he felt much better. May 23d. the cough was much better. The left side was dull at the upper part, and a dry creaking was replacing the crepitant rale. He was ordered five grains of the hypophosphite of lime, in place of the soda salt. May 30th, he was much amended, and there was but little sputum. June 13th, he felt himself well."

"B. D., 35 years old, was seen in June. He had had a bad cough since March, with frequent spitting of blood. The bowels were inclined to diarrhœa. Both sides of the chest were somewhat flattened. The respiratory sound was weak. Crepitant rales, to a slight extent, were heard over the left upper third of the chest. Cod-liver oil, he said, 'always ran through him.' He was ordered to take five grains of the hypophosphite of lime in the infusion of columbo, three times a day. He took this until August, when he was discharged, stating that he could now walk a long distance without fatigue; the cough also was 'nothing worth speaking of.'"

"R. H., 34 years old, had had a bad cough for the last five months. The sputum was often streaked with blood, and he sweated profusely at night. The respiration was weak at the left apex. From May 20th to June 5th he took five grains of the hypophosphite of lime three times a day, with five grains of Dovers powder at evening. In June he was let go at his own request, having lost entirely all cough and sweating."

The *preventive treatment* of pulmonary phthisis is a very important part of this subject. The condition of the system which favors the development of this affection (unless the doctrine that consumption is contagious is accepted) is in most cases hereditary; it may be acquired. Children born of either tuberculous parents, or of parents belonging to tuberculous families, have this constitutional tendency. These children demand *active, careful and persistent* attention throughout their infancy and childhood, their adolescence, and until they have passed the full meridian of adult life, if they are to escape the development of the malady.

The commoner manifestations of this scrofulous condition are glandular enlargements, which may suppurate and form foci from which the virus may be transferred to other parts of the body, where it will set up destructive changes. These children are especially liable to chronic inflammation of mucous membranes, and are frequently seen with inflamed eyes and eyelids, gastric disorders, diarrhœa on slight provocation, etc. They are also peculiarly susceptible to the deleterious influence of cold, and are apt to suffer from nasal catarrh and bronchitis, beginning in autumn and continuing throughout the entire winter. One of the most efficient remedies that can be prescribed in this class of cases is the *Hypophosphites.* Speaking of their use in the treatment of phthisis, Dr. BURNEY YEO, physician to the Brompton Hospital for Diseases of the Chest, London, England, says that "he has seen the greatest benefit in children in all forms of chronic lung disease. The patients felt stronger, were in better spirits, were more active, ate better, and slept better."

Children with a marked hereditary tendency to pulmonary phthisis can be kept beyond the middle period of adult life only by a rigid adherence to hygienic measures and a judicious medicinal treatment. To *prevent* the development of the disease should be the first aim of the parents and the physician.

More than two hundred thousand persons are now suffering in the United States alone from consumption and allied diseases of the tubercular type, of whom one hundred thousand will have died within the year, leaving their places to be filled by an equal number of future victims. Of the whole number of adults doomed to die in the prime of life, nearly one-third will be cut off by consumption.

Ancell, the author of one of the most learned and exhaustive treatises on the subject of "Tuberculosis," expresses himself as follows: "At a moderate calculation, according to the estimated population of the earth, and the almost universal prevalence of the disease, from eighty to one hundred millions of its present inhabitants will meet with a premature death from one form or other of it. It destroys, annually, nearly one-sixth of the population of this country, and comprises perhaps one-fourth of the practice of physic and surgery."

If by some sudden and unforeseen catastrophe, some dreadful epidemic and scourge like diphtheria, small pox, or yellow fever, one hundredth part of this number were destroyed each year, universal pity would be aroused, and the public mind would not rest until a remedy had been found. But disaster and suffering a hundred times more extensive, and a thousand times more ruinous to the commonweal, are acquiesced in with resignation or indifference because they are supposed to be *inevitable.* Yet if facts have any value, or the almost unvarying scientific testimony of medical experts any weight, no unprejudiced physician, no person of average education can, after reading this little pamphlet, come to any other conclusion than that a part of this premature death, this wasting of the life, the promise and the vigor of the country, might be stopped within a few months.

EFFECTS OF THE HYPOPHOSPHITES UPON THE SYSTEM.

"ONE of the first effects produced by the use of the hypophosphites is a general increase of nervous energy, with a feeling of ease and comfort. The second effect is an increase of appetite; digestion is improved, and the bowels become regular in their action. The quantity and color of the blood is increased, so that in females menstruation becomes easier, more abundant, and more regular; respiration is controlled, a better expansion of the chest is observed, cough improves, easy expectoration is produced, night perspiration diminishes, the face becomes fuller, the lips red, the nails and hair grow, and in children the teeth, showing the importance of the hypophosphites on the organ of nutrition."—*Dr. Taylor*, *Lancet.*

"In the present state of medical science," says Dr. Churchill (p. 50), "phthisis, when not treated by the hypophosphites, at whatever period of development it may be observed, however early the stage at which it may be taken in hand, must be regarded as almost always fatal.

"All the means hitherto used have no certain action except

against the accidental complications of the disease; its essential causal conditions remain completely beyond their control, for under the most able and most experienced hands the number of recoveries is *less than four per cent.*

"In the very rare cases where phthisis ends in recovery, it has been hitherto impossible to determine what were the conditions to which this result was due. As the prognosis of the almost certain fatality of phthisis depends solely on the *nature of the disease itself*, it is impossible to determine beforehand, with the least probability, whether any given case will be an exception to the general rule.

"On the other hand, when phthisis is submitted to the specific treatment of the hypophosphites, the prognosis may, in the great majority of cases, be established with a degree of certainty equal, if not superior, to that of any other curable disease.

"It rests upon two series of conditions: the extent of existing pulmonary lesion, and the presence or absence of complications. These may be summed up as follows:—

"When there are no complications, the prognosis may be thus stated:—

"1. Phthisis in the first stage ends in recovery.

"2. It also ends in recovery in the second stage, provided one lung only is affected.

"Hence it follows that the use of the hypophosphites is followed by recovery in cases where the local disease has not proceeded beyond certain limits, and that consequently every patient *may be cured provided the hypophosphites be used in time.*

"3. Phthisis in the third stage, when limited to one lung, may also be followed by a cure.

"4. When tuberculosis in the second stage exists in both lungs, recovery depends upon two conditions. First, the cessation of the diathesis, which is brought about by the proper use of the hypophosphites; and, secondly, upon the possibility of arresting, softening, or compelling it to proceed but slowly, which may be frequently obtained by appropriate secondary means. The favorable conditions in such cases are when the deposit is only partial in both lungs, or when, if diffused throughout both lungs, the tubercles are disseminated and far apart.

"5. When the disease has reached the third stage, and tuberculosis has attacked both lungs, recovery is still possible.

"6. When there are cavities in both lungs, recovery takes place in a few exceptional cases.

"I have seen three examples of this. One was a lady who had treated *herself* by the hypophosphites, and who after her recovery came to consult me because still suffering from dyspnœa, without any other symptom. She was very stout. The other two cases which were under my treatment are both in perfect health. One is an officer in the Imperial Guard, and has been doing duty for two years past.

"7. The prognosis of acute phthisis rests on the same principles as that of chronic, but it offers more uncertainty on account of the difficulty of ascertaining the extent of the already existing lesion, and also of distinguishing between acute tuberculosis and acute softening.

"8. In children, the prognosis is much more favorable than in adults, which is altogether contrary to what is observed with any other mode of treatment. In children I have seen recovery take place with the hypophosphites, in nine cases out of ten, at all stages of the complaint.

"9. With the hypophosphites, patients who have an hereditary predisposition have more chance of recovery than those who have no predisposition.

"Within the last ten years there has been an appreciable falling off in the mortality of consumption. That the use of this remedy is the real cause of this decrease is shown by Dr. Bennett confessing that he has given the hypophosphites to a *large proportion* of his patients, and that nearly fifty per cent of those so treated had got well. Dr. Williams acknowledges that, with cod-liver oil alone, previous to 1862, he had not cured two per cent of his patients; that when the oil fails, he finds the hypophosphites to succeed, and that since 1862 (*i. e.*, since the introduction of the hypophosphites) he has cured seventy-five per cent. Drs. Williams and Bennett used neither phosphorus nor the phosphates; they used the hypophosphites, and there is not upon record a case of the cure of consumption by free phosphorus."—*Churchill*, p. 380.

"Were I only to quote the successful cases that I have had under my care, the cases in which the lung disease has been arrested, and even cured, *I would quote many instances of cure, myself included*, which have apparently taken place under the influence of the hypophosphites as they were long and constantly administered."—*Dr. Henry Bennett*, *Pulmonary Consumption*, p. 99, 1871.

Dr. J. C. Brown, lecturer on mental diseases to the Leeds School of Medicine, says: "We owe a debt of gratitude to Dr. Radcliffe for pointing out the value of the hypophosphites in debility and nervous diseases."

J. Thorowgood, M. D., Physician to Hospital for Diseases of the Chest, Victoria Park, England, says: "I have administered hypophosphites with a view to increasing and restoring wasted nerve force, in an exhausted state of which I believe many cases of consumption to have their origin. Such experience as I have had leads me to think, with some others who have made careful trial of the hypophosphites in consumption, that these salts are certainly to be regarded as valuable remedial agents in the treatment of this disease, especially in its premonitory and early stages."

Dr. Lieto Regnoli, Physician-in-Chief to the Hospitals, writes to Dr. Fideli as follows: "For several years past I have adopted in my private practice, for the treatment of pulmonary tuberculosis, Dr. Churchill's chemically pure hypophosphites, and I have endeavored scrupulously to carry out his indications. I have treated in all *twenty-one cases* in different stages of the complaint. Of these, seven in an advanced state died; nine, *who were in the condition of curability* indicated by Dr. Churchill, were cured; the other five were very much improved. Among those who recovered, I was much struck by the following:

"CASE I.

"A girl of eighteen had come to Rome from the country with all the symptoms of advanced phthisis in the second stage; there was extreme emaciation, loss of appetite, cheeks highly colored, loss of voice, burning heat, fever in the evening, profuse perspiration in the morning, troublesome cough, with thick purulent expectoration, menstruation suppressed, etc. The use of the hypophosphites alone, for three months, was followed by complete recovery. On her leaving Rome, she was directed to continue the hypophosphites from time to time. Since then three years have gone by. She has married, has become a mother, and continues to enjoy good health.

"CASE II.

"A consumptive lady in the fourth month of her pregnancy and suffering from consumption in the second and *third* stages was cured of the complaint by the use of the hypophosphites; but independently of the recovery of the mother, there was a marked improvement in the constitution of the child as compared with those she had had before. This fact particularly struck me, and induced me to follow up the experiment.

"CASE III.

"Another lady, who was not consumptive, but rachitical, who had previously had four scrofulous, rachitical children that had died in their infancy, was ordered during her pregnancy to take a tablespoonful of —— Syrup daily. The result was that she gave birth to a well formed, healthy boy, who was suckled by his mother, and grew up so healthy and strong as to look as if he belonged neither to the mother nor any of the family. In another pregnancy the same lady was put under the same treatment. This child was also a fine, healthy son. He was suckled by his mother, who took the hypophosphites from time to time. The boy is now in perfect health, and three years old."

Since then Dr. Regnoli has again written to Dr. Fideli to say that other successful cases since observed had confirmed him in his first opinion, that the hypophosphite treatment is a most valuable means of obtaining successful results in the treatment of consumption, providing a CHEMICALLY PURE *hypophosphite can be obtained.*

OTHER USES.—The great reputation which the hypophosphites of lime and soda have acquired, has been founded chiefly upon the good results obtained by their use in the treatment of pulmonary phthisis. The accepted facts of materia medica and therapeutics, however, justify us in expecting from them equal benefit in the treatment of other affections.

Phosphorus has an undisputed value in the treatment of a large number of diseases. Lime and soda, in their many chemical combinations, have occupied a prominent place among therapeutic agents from the earliest use of medicines.

Experiments upon animals have proved that phosphorus may be a tissue food. It exercises a specific formative action on bone tissue. It accelerates the ossification of temporary cartilages. It can be taken without producing any deleterious effects whatever on either the general nutrition or any particular organ. It, therefore, has been, and still remains, the foremost among the remedies used in the treatment of rickets, and some other diseases of the osseous system. The benefits which it produces in this class of cases is unquestioned.

Phosphorus, also, has a value in the treatment of diseases of the nervous system, which is not less than that claimed for it in the treatment of bone affections. Hence, compounds of phosphorus have been used with great satisfaction in the treatment of brain exhaustion, neurasthœnia and functional impairment of memory. It is clearly indicated in all cases of functional disorder, which depend upon lack of nerve force, such as sexual debility, actual impotence and spermatorrhœa. It cannot be expected of any remedy that it will do everything in this latter class of cases, and it is indispensable that all vicious habits be stopped; when this has been done the compounds of phosphorus, especially, the hypophosphites of lime and soda, will impart the tone and energy to the nervous system essential to complete restoration to health.

Besides, it has been satisfactorily shown that the compounds of phosphorus exert an important influence on the construction of the red blood-corpuscles. It was thought, at one time, that this power would make it a valuable remedy in the treatment of *leucocythæmia*, but no results have been obtained which justified this hope. But that phosphorus and its compounds *do* exert a beneficial influence upon the corpuscular elements of the blood in all conditions attended by *anæmia* has been fully established. Hence, these remedies are of great service in the general debility which follows exhausting diseases, such as fevers and small-pox, cholera infantum, scarlet fever, severe chicken-pox, complicated and exhausting vaccination, and any wasting disease of childhood; weakness and anæmia, produced by abscesses, boils, carbuncles; and also that impoverished condition of the general system which frequently attends the severer diseases of the skin, as pemphigus, suppurating eczema, etc.

In the healthiest human body there is the completest balance between waste and supply, and the highest degree of nutrition is maintained when the equilibrium between waste, supply, and all the demands of growth and function is most perfect. It is to this end that all remedies are administered; to correct impaired digestion, to supply the deficiencies existing in different tissues and organs; to repair the waste incident to disease, indiscretion and increasing age, etc. The medicinal agents now under consideration have long been used for these purposes, either alone or in combination with each other, and also mixed with other remedies (see formulas, pages 47 to 51) designed as adjuvants. To be of the greatest service, however, the hypophosphites must be taken in large quantities, and continuously for weeks or months. This is rendered necessary by the fact that when they are put into the alimentary canal in small quantities they are very liable to be swept away by the products of the digestive track. This is especially true when the salts themselves are taken, or are taken in solutions, such as are ordinarily prepared. When taken freely the best results will be obtained. To secure a combination which would preserve them in the best possible condition to be absorbed, and with the least liability to become inert on standing, has been one of the leading objects of my chemical experimentation concerning these drugs. We have ascertained that in the form of syrup made from *beet sugar*, by cold maceration, we have the only form in which they can be kept free from decomposition, and therefore the only preparation in which they do not become inert.

This has been the great obstacle which has heretofore stood in the way of a more extensive use of the hypophosphites, and now that it has been effectually overcome, these remedies can be prescribed with as much certainty that they will produce effects as any class of remedies in use in medicine.

SELECT MEDICAL OPINIONS.

1 EAST 42ND ST., NEW YORK, Jan. 7, 1884.

DEAR DOCTOR:

I have intended writing you many times to add my mite in the praise of your valuable preparation of hypophosphites, but have been prevented in one way or another from doing so. There is no medicine that I prescribe more frequently and more successfully than your preparation of hypophosphites. I have used it in tubercular infiltration of the lung many, many times, with the happiest results, and have seen such processes stop, then recede, and finally leave no trace of any prior morbid condition. I have also used it in patients convalescing from croupous and broncho-pneumonia, where the exudation is slowly or not at all absorbed, and in such cases if this medicine be used *persistently* for *many weeks* all morbid conditions disappear, and the lung is free once more to carry on its vital function.

I think one reason why physicians do not always succeed with these hypophosphites is that they do not instruct their patients to take them *long enough*. In neurasthenia and allied nervous disorders, these hypophosphites may be used with great advantage often.

Respectfully,

J. MONTFORT SCHLEY, M. D.

Professor Physical Diagnosis, Woman's Medical College, &c.

Referring to Dr. Schley's opinion, that the reason why physicians do not always succeed with these hypophosphites is that *they do not instruct their patients to take them long enough*, I beg to say that every organic modification in the animal system requires *time*. The process of repair takes place but slowly, and, we may say, by insensible steps.

Robin (Robin et Verdeil, "Chimie anatomique et physiologique," Paris,) says: The cure of a general disease requires a *long process of nutrition*, and for the *time* thus required no substitute can be found. We may foster it, hasten it, prevent it from going back; but we cannot suddenly restore the elements of the diseased substance to their first state of molecular combination and relative quantity, etc. For all this *time* is required. Thus the successive study of the different parts of the system gradually destroys the chimerical hopes which misled our predecessors, such as the discovery of universal remedies, and in lieu of them gives us the notion of power, limited on our part, but really existing. We must, therefore, give up our vague aspirations towards unlimited and indeterminate power, and for it we must substitute the idea of attaining, under certain conditions of time and proper means, a real result which has been determined beforehand."

The following is an extract from a letter just received from C. Irving Fisher, M. D., late Port Physician of Boston:—

"On July 30, a young man presented himself at my office with the following history: Age, thirty; trade, laster of boots; mother living, but very feeble with 'old-fashioned consumption;' one brother died with the disease; for a long time had been losing flesh and strength; had an annoying cough, with pain in left side of chest, and night sweats. Examination of the lungs showed that in the upper portion of the left lung phthisis was well established. I prescribed

R. SYR. HYPOPHOS: COMP: McARTHUR

alone. After taking the syrup a few weeks, I have rarely seen a patient so enthusiastic regarding his improved condition. The cough had nearly disappeared, night sweats completely so, appetite improved, flesh and strength returning, etc. I saw him the first of January, four months after beginning treatment, and learned that he continued to improve, and had regained his usual health."

HOLBROOK, MASS., March 21, 1883.

DEAR MCARTHUR:

I received the pamphlet and I like it much. Since you were here I have prescribed it a good deal as a tonic with liq. strych., iron, etc.; once with ergot and sulph. atropia for incontinent urine, and with every promise of a splendid result. The case is of long standing, and has not before yielded to ergot and atropia in other combination.

Yours,

C. I. FISHER, M. D.

John Dixwell, M. D., General Agent Massachusetts Society for Lost, Stolen and Abused Children, writes:—

BOSTON, May 21, 1879.

DOCTOR MCARTHUR:

My Dear Doctor,—After having given your Syrup of Pure Hypophosphites a fair trial in some dozen typical cases of pulmonary trouble in widely differing subjects and in various stages of advance in disease, I feel prepared to express an opinion of some possible value as to the action of this preparation.

I have noticed *distinct curative powers* in the *first* and *second* stages of phthisis, male and female adult, and in first stages of children over twelve years of age. In some female cases noticed an increase in the menses, apparently as thus caused. Even in the *third* stage of phthisis, in three cases of adults, an evident prolonging of life by supply of this nutritive element was recorded. Hence, I want to add my voice to others in favoring the use of your preparation when symptoms demonstrate the need of a powerful, easily assimilated element of this sort.

Very truly, etc.,

JOHN DIXWELL, M. D.

John Dixwell, M.D., General Agent Massachusetts Society for Lost, Stolen and Abused Children: —

BOSTON, MASS., Aug. 29, 1880.

MY DEAR DOCTOR:

Let me say that your Hypophosphites have done more for three little broken-down children, who have come under our care of late, more than we ever expected from our use of the preparation before. We do not want better results.

LAKE FOREST, ILL., July 2, 1879.

For forty-four years I have suffered with dyspepsia, being able to keep alive only with the greatest carefulness in my hygienic and physiological habits. My alimentary organs are the weakest and most sensitive of any person I ever knew, therefore my effort has been to keep up by the most nutritious and easily digested food. In addition to ordinary food, I have made use of extracts containing phosphates and nitrates. Have also used Liebig's extract meat, but it has produced colic; have used extract wheat with malt, which produced bleeding piles, so had to lay it aside; then used *vitalized phosphates* with a like result. Observing that your hypophosphites was made with beet or French sugar, I began six weeks ago to take them, and they have improved my general health and strength more than anything else I have tried, and without causing any local disturbances whatever. I do not believe this would have been the case if you had used American sugar, or if the hypophosphites had not been *chemically pure.* The coffee sugar being refined with alum, bisulphite of lime, etc., produces dyspepsia and sore protruding piles, while the powdered sugar, being refined with acetate of lead, causes bleeding piles, etc.

I think some eighteen persons in this place have of late been experimenting with your Syr: Hypophos: Comp: with most decided benefit.

Most cordially yours,

LUTHER ROSSITER, M. D.

J. A. MCARTHUR, Lynn, Mass.

"I have frequently prescribed Dr. McArthur's Syrup of Hypophosphites, and have found it to be a therapeutic agent of great value in the treatment of many diseases."

O. G. Cilley, M. D., Surgeon-General, Mass.

Baltimore, March 22, 1881.

Dr. J. A. McArthur:

My Dear Doctor,—I take this occasion to say that I used the Syrup of Hypophosphites you were kind enough to furnish me for trial in suitable cases in the City Hospital, both in the in- and out-patients' departments, and in all cases with very decided benefit. In every case emaciation was arrested, and in some there was a decided gain of flesh and strength, with a corresponding improvement in the cough and other symptoms.

I now prescribe it habitually in my private practice, and ***always with benefit*** when the cases are properly selected. Of course, when patients come to me with cavities already formed, and other points ready to soften and break down, in a state of extreme emaciation, and suffering from pronounced hectic, I do not prescribe your Syrup with any expectation of permanent benefit. But, even in these cases, I think the patient's condition is improved, waste is retarded, and ***life thus prolonged.***

I am your most obliged and obedient servant,

John S. Lynch, M. D.,

Prof. Principles and Practices of Medicine and Clinical Professor of Heart, Throat and Lungs, in College of Physicians and Surgeons.

"I have used Syrup Hypophos: Comp: McArthur for four years with most excellent results. It is very valuable in phthisis pulmonalis, bronchitis and many forms of infantile diseases, especially when the osseous system is defective."

Gertrude G. Bishop, M. D.,

310 Throop Ave., Brooklyn, N. Y.

SALEM, MASS., NOV. 22, 1877.

MY DEAR DR. MCARTHUR:

Let me again thank you for the Syrup Hypophosphites that you kindly sent me. It is very, very seldom that I seek aid from medicines other than those that have their place in the "Homœopathic Materia Medica," after thorough "*proving;*" but my friendship for yourself and my confidence in your statements as to the benefit that had resulted from the use of your Syrup, caused me to give it a trial. Two patients to whom I gave it certainly derived great benefit from its use. One of them was thus spared the trouble and expense of a voyage to a distant island, and seems now in perfect health.

Sincerely yours,

SAM'L H. WORCESTER, M. D.

UTICA, N. Y., January 1, 1884.

J. A. MCARTHUR, M. D.:

Dear Doctor,— For several months I have been using your Syr. Hypo. Comp., and so well pleased am I with its effects, that I have no hesitation in recommending it to physicians as a *pure* and *reliable* preparation.

It is also very palatable, and never disturbs the most delicate stomach. I have called the attention of several physicians to it, and, after giving it a fair trial, they all bear testimony to the excellent results obtained. It is now kept on sale by our principal drug stores, and I use no other hypophosphites.

Trusting you may have success in placing before the medical profession a reliable and honest medicine, I am,

Truly yours,

C. B. TEFFT, M. D.

BOSTON, Dec. 16, 1882.

DEAR DOCTOR:

I have used your Syrup of Hypophosphites in many cases of phthisis, and as a rule with good results. In cases where cod-liver oil disagrees, as it often does, especially where the appetite is poor and the digestive powers much weakened, I have found it a welcome substitute to patients who loathe oil in any and every form. It is something more than a placebo. In cases not too far advanced its value is soon apparent. I have more than once noticed that where the usual remedies against night sweats have been powerless, the Syr. of Hypophosphites has acted as a powerful auxiliary, speedily checking this exhausting and troublesome symptom. It seems also to diminish the cough. I am sure it increases the appetite and improves digestion; hence I have frequently given it to feeble and ill-nourished children, and almost invariably with excellent results.

Very truly yours,

PRINCE W. PAGE, M.D.

"I wish to give my testimony to the value of Dr. McArthur's *Syrup of Hypophosphites.* I have used it and am *now* giving it in what I call a typical case of phthisis with a most excellent result. The patient (a young lady belonging to a phthisical family) has improved in every way, has gained in flesh and strength, cough much less severe, night sweats gone, catamenia returned, etc. I had previously, and for months, used every remedy I could devise without any benefit. She is now taking the fourth bottle. I believe the value of the syrup cannot be over-rated, as it supplies the elements of nerve nutrition in an easily assimilable form."

L. F. WARNER, M.D.

Hotel Pelham, Boston, Mass.

PEABODY, MASS., Dec. 10.

DR. J. A. MCARTHUR:

Dear Sir,—I have used McArthur's Syrup Hypophos: Comp: quite extensively in the last few years, and from the satisfactory results obtained I shall use it more frequently. In the case of my servant girl, who, to all appearances, was strongly tending to consumption, with loss of appetite, loss of flesh, a short, hacking cough, sharp pains through her lungs and a pallid complexion, the effort to perform any work was attended by almost complete exhaustion. I put her on your preparation of Hypophosphites and stopped all other medication, and after taking the second bottle she expressed herself all well, and a month after again took service in a large family and feels nicely. I have used it with gratifying results in cases of marasmus, and in the case of my own baby during the excitement incident upon teething, and in cases for which it is intended, it has fully met my expectations.

Respectfully,

F. S. WORCESTER, M.D.

DR. E. H. SPOONER, M. D., 800 DeKalb Avenue, Brooklyn, N. Y.:

"For more than a year I have used McArthur's Syrup Comp: Hypophos: with the best results. I recall now six cases of phthisis which I have treated during the past year, which have *been cured by the use of the Syrup, and several of these cases were in the second stage of the disease, etc.*"

BALTIMORE, MD., Sept. 28, 1882.

DR. J. A. MCARTHUR:

Dear Sir,—I have been using your Syr: Hypophos: Comp: quite extensively in practice for two or three years, and with excellent results.

Yours very truly,

J. L. INGLE, M. D.,
247 Lauvale Street.

MARENGO, WAYNE CO., N. Y., June 1, 1883.

J. A. MCARTHUR, M. D.:

Dear Doctor,—I am a very strong believer in your treatment of consumption, and I believe that most of the cases can be cured if we can get pure remedies. I have of late been using your Syrup Hypophos: Comp: and find it pure. Since I have been using your syrup my patients recover.

Yours truly,

L. W. SUTHLAND, M. D.

GRAND RAPIDS, MICHIGAN.

Please send me your pamphlet on "Consumption and Tuberculosis." I am using your Syr: Hypophos: Comp: in "paralysis agitans" of some ten years' standing. I have used it with advantage in the general prostration following spinal meningitis. I am satisfied it is more than an ordinarily good preparation.

Yours respectfully,

ISAAC WATTS, M. D.

486 South Division Street.

BALTIMORE, Oct. 16, 1884.

J. A. MCARTHUR, M. D.:

Dear Doctor,—I have prescribed your valuable preparation of Hypophosphites in many cases of tubercular infiltration of the lungs, with good results, and have seen such process stopped and recede, leaving no trace of a prior morbid condition. I am satisfied that a trial of the syrup as made by you is all that is required to establish its therapeutic value.

Very respectfully yours,

J. A. GILLISS, M. D.

150 North Eutaw Street.

BOSTON, Jan. 14, 1884.

MY DEAR DOCTOR MCARTHUR:

I have been using your Syrup Hypophos. Comp. for several years in cases where the Hypophosphites seemed to be indicated, and have been much pleased with the results obtained; so much so, that I am now in the habit of specifying McArthur's.

I would be pleased to see you or to hear from you at any time.

I am yours very truly,

H. M. JERNEGAN, M. D.

700 Tremont Street.

DR. CHARLES AHEARN, of Lynn, graduate of Harvard, writes: "I have used Dr. McArthur's Syrup Hypophosphites Comp. for the last three years, with increasing confidence in its efficacy, not only in cases of incipient phthisis and bronchial affections, but also in difficult dentition and affections of the brain and nervous diseases."

From F. LE SIEUR, M. D.,

Lecturer on Dermatology, Medico-Chirurgical College, Philadelphia.

PHILADELPHIA, June 8, 1885.

DEAR DOCTOR MCARTHUR:

In cases of skin disease, where the system is far below the normal standard, with progressive waste of tissue, a remedy which can arrest this morbific process, and, indeed, re-supply the waste, is pre-eminently indicated. Such a remedy I find constant use for in my clinic at the Medico-Chirurgical College, and among the many preparations the Hypophosphites rank superior in many instances to cod liver oil. In this connection, I want to tell you, my dear doctor, that your Syrup of the Hypophosphites has given me the most unbounded satisfaction, so satisfactory that I deem it no more than proper and just to inform you of the fact. May success attend you.

Faithfully yours,

F. LE SIEUR.

BALTIMORE, MD., Sept. 7, 1882.

"I take this early opportunity to write in order that I may contribute any aid I can towards bringing your meritorious preparation into the prominence it is entitled to occupy as an alterative. You are at liberty to say that after two years' use of Dr. McArthur's Chemically Pure Syrup of the Hypophosphites in my practice, I am prepared to reiterate with emphasis what I said in a medical journal at that time, viz.: 'I do not hesitate to commend it to the profession as worthy of confidence.'"

PROF. HARVEY L. BYRD.*

BALTIMORE, April 26, 1881.

DR. J. A. MCARTHUR:

Dear Doctor,—You are at liberty to use my letter in any way that will contribute to your pleasure or profit. There are so many nearly worthless proprietary medicines of this kind on the market (and every day is adding to their number) that when one is found to be really valuable I think it the duty of the profession to make it known.

JOHN S. LYNCH.

"I find McArthur's Syr: Hypo: Comp: to excel any other preparation of the kind I have ever used. I have ordered it in this city and in Brooklyn, N. Y., and to my patients as far west as Dakota, with the very best results. * * * I am perfectly delighted with it."

JOHN CALDWELL, M. D, Baltimore.

---

* The death of Dr. H. L. Byrd occured at Baltimore November 29th. He was formerly a resident of Mobile. His record as a successful physician, a surgeon in the Confederate Army, and a popular medical editor, has caused his name to be well and widely known.—*Louisville Medical News.*

HYDE PARK, MASS., Sept. 13, 1885.

DEAR DOCTOR:

I thank you cordially and sincerely for your courtesy in sending me your circular, which I have read with interest and benefit, and as one who is *especially* interested in the treatment of "*Children's Diseases.*" I shall always be glad to hear from you, as I have learned by experience to value your Syrup Hypophosphites in the treatment of that class of cases. A reliable, trustworthy preparation of that character is a valuable adjunct to remedial treatment, and in no class of cases more so than among the children. If I can be of service to you in any way, please command my best efforts.

Yours very truly,

CHAS. STURTEVANT, M. D.

BROOKLYN, N. Y., Sept. 25, 1885.

DR. J. A. MCARTHUR:

Dear Sir,—I have for a number of years past prescribed your *Syr: Hypophos: Comp:* with much satisfaction.

Yours very truly,

E. W. OWEN, M. D.

LYNN, Jan. 14, 1886.

DR. MCARTHUR:

Dear Sir,—For three years I have used your Syrup of the Hypophosphites. At present I use none other, as I have found by experience that it is a very reliable, elegant and satisfactory preparation; excelling, in my opinion, all others in the market. I cordially and conscientiously recommend it to the profession.

F. D. STEVENS, City Physician.

CHICAGO, June 27, 1886.

J. A. MCARTHUR, M. D., Lynn, Mass.:

Dear Doctor,—Your *Syr: Hypophos: Comp:* is a valuable remedy, whose uses are fairly set forth in your little pamphlet.

Yours truly,

C. C. HIGGINS, A. M., M. D.,

Med. Director Chicago Guaranty Fund Life Society.

ST. ELMO, ILLS., July 3, 1886.

J. A. MCARTHUR, M. D.:

Dear Sir,—I have been using your Syrup Hypophos: Comp: for several years, and have perfect confidence in its efficacy. In a variety of diseases, it will do all that is claimed for it.

Respectfully,

D. F. LANE, M. D.

REPORT OF CASE BY DR. McARTHUR.

FOR obvious reasons, I have refrained from reporting any cases treated by myself with the Syrup Hypophosphites; but, as the following so nearly concerns me, I venture to copy it from my note-book:—

Patient, my brother, E. R. McArthur, aged thirty-eight years; residence, Marblehead; December, 1875.—Has had a cough and been ailing for several months, with muco-purulent expectoration, great debility, and considerable emaciation, with excessive night-sweats. Has had several attacks of hæmoptysis; one very severe; thinks a pint or more of blood. His voice is hoarse, and has tickling in throat. Suffers from dyspnœa, and is obliged to desist from business; no appetite, digestion bad, vomiting frequent, and some diarrhœa. Had taken various emulsions of cod-liver oil, malts, tonics, etc., but without finding any relief.

Local Signs.—Expansion of two sides of thorax unequal, the left side hardly rising at all during inspiration. Front, dulness on percussion over the whole left side; diminished resonance on the right side for three fingers' breadth below clavicle. Auscultation revealed on left, above clavicle, blowing respiration, with gurgling, and moist crepitus; cavernous cough and pectoriloquy; below clavicle, same signs, *i. e.*, moist cracklings extending down to base. Above clavicle, on right, jerking respiration, with moist rales, increased vocal resonance, exaggerated respiration in the rest of lungs. Behind, on percussion, dulness in left supra-spinous fossa. On auscultation, on left, sharp cracklings in the infra-spinous fossa.

Diagnosis.—Cavity at apex of left lung, with tubercles over the whole of its front aspect; also softening at front of apex of right lung.

I prescribed:

R. SYR: HYPOPHOS: COMP: MCARTHUR.

Dose.—A dessertspoonful, gradually increased to a tablespoonful, after each meal, in a little port wine and water.

Feb. 1.—Has begun to improve; cough somewhat

diminished, less perspiration at night, and better digestion. Continue a tablespoonful of the Syrup three times daily, with liberal diet and wine.

Feb. 15.—Improvement continues.

March 1.—Stethoscopic signs considerably improved; less dulness and fewer rales. Continue treatment.

March 20.—Manifest improvement; has gained seven pounds in weight; feels better and stronger than at any time since the commencement of his illness. As stomach and digestion were now good, ordered cod-liver oil, a tablespoonful two or three times daily, one hour after the meal, and continue the Hypophosphites in whiskey and water.

May 1.—General condition greatly improved; slight cough in morning, with a little expectoration; appetite good; voice recovered its natural tone. Omit Syr. Hypophosphites for a time and take liberal diet, with porter or ale.

June 9.—Reports cough ceased almost entirely; no expectoration, has gained flesh and sleeps well; no thoracic pains or difficulty in lying down; no fever or night-sweats; hoarseness has disappeared; can sing as well as ever; appetite good, and can attend to his business as well as before his sickness.

He took the Syrup at intervals for the next two years, as he said he felt better by doing so. I saw this patient about one hour ago (Oct. 1, 1886). He continues to enjoy perfect health, and says, "*I never felt better in my life.*" He is at the present time in *charge* of my laboratory.

In connection with this case, I wish to call the attention of physicians to one very important consideration as to the therapeutical action of the different hypophosphites. I have experimented with a number of these salts, particularly those of soda, lime, iron, potash, manganese, etc., and am convinced that in the *treatment of phthisis the hypophosphites of lime and soda* should alone be used.

Ten years ago Dr. Churchill wrote: "Extended experience has now shown me that in the treatment of tuberculosis the practitioner should confine himself to the use of *lime* and *soda*. The effects produced by these two salts, when properly combined and administered in phthisis, have such a character of constancy, that I have seldom thought myself justified in intermitting them or supplying their place by any other." He further says: "In the first edition of my work on Consumption, I stated that the action of the hypophosphite of iron should only be tried with *great caution* in cases of consumption, as in several for which I had prescribed ferruginous preparations simultaneously with the hypophosphites, their exhibition appeared to be followed by hæmoptysis, or hemorrhage in some shape or other. Subsequent experience has since fully confirmed these views. Not only was hemorrhage produced in almost every instance in which I used the hypophosphites of iron, but in patients who have been previously taking ferruginous medicines, it will be found that it is very difficult, at first, to keep the effects of the hypophosphites within the limits of their *physiogenic* action. For this reason, I have entirely given up the use of the hypophosphite of iron in phthisis, and confine myself almost entirely to the use of the CHEMICALLY PURE SYRUP of *Lime and Soda*."

My own observations entirely coincide with Dr. Churchill's remarks, and it was during the administration of a well-known Compound Syr. of Hypophosphites, which I subsequently learned contained iron, manganese, strychnine, etc., that my brother had his first attack of hæmoptysis, and subsequent observation has convinced me of the highly dangerous character of these compounds in phthisis when there is the least tendency to a hemorrhagic diathesis. I could copy from my case-book many instances to verify the above, cases that have occurred, not only in my own practice, but in the service of other physicians with whom I have been called to consult. A case in point occurred but a few months ago, and is briefly as follows: I was called in consultation to see a young lady who was suffering from, not profuse, but nevertheless troublesome and persistent hæmoptysis which had lasted a number of days, and could not be controlled by any of the usual remedies. On inquiry, I found that she still continued to take three times daily a Compound Syrup of Hypophosphites, which on examination proved to be an impure article, containing among other drugs, iron, strychniæ, etc. This medicine was ordered to be discontinued, and in two days the hemorrhage ceased. At the end of one week she was ordered a dessertspoonful of McArthur's Syrup Hypophosphite of Lime and Soda in port wine and water, three times daily, with the meal; one tablespoonful of cod-liver oil one hour after breakfast, and tea, and liberal diet. There was no more bleeding, and she is now making a good recovery.

OLD AGE.—Dr. Dawson, in a recent number of *Knowledge*, has discussed this subject with much skill, and other writers have established some very interesting facts. Research has demonstrated that the principal anatomical characteristics of old age is a deposition of fibrous gelatinous and earthy deposits in the system, and that the *symptoms* of *mental* decay resemble the gradual change that comes over old people. It is proved that a man grows older and mentally weaker as the brain contains less phosphorus. Protogen, an important constituent of the brain, is largely made up of phosphorus, and foods richest in that element will restore most speedily weakened brain power. "The pathological phenomena discovered in the brains of persons dying insane all have for their basis interference with the due nutrition, growth and renovation of the nerve cells, which, by interrupting the nutrition, stimulation and repose of the brain, essential to mental health, results in the impress of a pathological state in the brain, and disordered mental function."—DR. MANN.

We find that the blood contains compounds of lime, magnesia and iron, and as blood is made from the assimilation of food, it is to the food we must look for the origin of those earthy deposits which give us the "senile dementia and the decrepitude of old age." The following articles of food contain least of the earth salts: fruits, fish and poultry, young mutton and veal. Old beef and mutton contain much earthy matter.

As old age and mental decay approaches, the necessity of physical activity diminishes, and we therefore select the food containing the least amount of earthy matter, endeavoring at the same time to counteract the excessive amount of atmospheric oxygen, and as far as possible dissolve already formed calcareous concretions.

DR. DELACY EVANS shows that water bubbling up pure and clear from certain springs with dilute phosphoric acid, or what is better—*Hypophosphites*—possesses curative powers to a remarkable extent. The solvent properties of perfectly pure water has a powerful action in dissolving and eliminating those earthy salts which are blocking up the great blood vessels, and preventing their undue deposit.

The Hypophosphites—"McArthur's Syrup"—by its wonderful affinity for oxygen checks the gelatinous deposits, and prevents the accumulation of earthy salts. A writer in the *Medical Times*, referring to the subject, says: "To retard old age and mental decay, keep the mind fresh and vigorous to the last. As the sun of our life turns the meridian and descends towards its setting, keep up in a less degree than formerly, but still keep up, some active physical and mental labor, avoid food rich in the earthy salts, and take daily three tumblers of pure water—*Poland Spring*, for instance—adding to each glassful two teaspoonfuls of McArthur's Syrup, and, if preferred, the 'squeeze of a lemon.' The brain will be clearer, exertion less painful and difficult, and the close of life come more like the sun setting in a golden brightness, than as we often see in clouds and darkness."

DR. MCARTHUR'S Hypophosphites may be given also with benefit in other diseases than consumption, particularly those characterized by loss of nervous power.

DR. REGNOLI cured several cases of paralysis, dependent upon deficient nerve power, in five or six months, by the exclusive use of the hypophosphites; he also uses it in the convalescence following diphtheria and fevers.

DR. VALENTINE recommends them in chronic bronchitis, and says with the syrup he has made some SPLENDID CURES.

DR. FEDELI uses them in scrofula, cerebral and mesenteric tuberculosis, in unhealthy and rickety children or infants, as well as in cases of amenorrhœa, anæmia, and adds: "The use of them has been followed by VICTORIOUS RESULTS, but the treatment must be kept up for a sufficient length of time, and directed with all possible care." He concludes by saying:—

"I must confirm what has already been stated, as to the ABSOLUTE NECESSITY of the hypophosphites BEING PERFECTLY PURE, and prepared according to the method that Dr. Churchill alone has communicated; for I find many preparations are too apt to do harm instead of good; hence, many of the failures which bring discredit upon the treatment."

DR. TANNER recommends this syrup in cases of extreme debility and mental depression, in rickets, scrofula, tabes mesenterica, in progressive, locomotor ataxy, epilepsy, hysteria, neuralgia, and in over-exertion of the brain; in nervous excitement, when sleep will not come, they act better and safer than opiates.

DR. WILLIAMS says they are invaluable in treating spermatorrhœa, uterine catarrh, and weakness during pregnancy and lactation.

DR. McCORMICK says "In sleeplessness, restlessness and difficult dentition of infants, and in all diseases of debility, the PURE hypophosphites are of paramount importance."

In intermitting and feeble action of the heart, combined with fld. extract digitalis, they have no equal.

PURITY OF THE HYPOPHOSPHITES INDISPENSABLE TO SUCCESS.

To show the difficulty of obtaining pure hypophosphites, and consequently of arriving at satisfactory results, I beg to give the following passages : —

"The point of primary importance in the use of the hypophosphites is their *chemical* purity, but, unfortunately, like nearly all substances employed in medicine, they are too often adulterated."

"I have met with salts containing carbonate of soda, free soda, sulphuret of sodium, which have been sold as pure hypophosphites."

"Of ten samples of hypophosphites of soda, lime, and potash, obtained from different manufacturers, and submitted to qualitative analysis, I found only three pure. The other seven contained either carbonates, phosphates, or free lime."

"When I stated years ago that the non-success of some practitioners in the treatment of phthisis with the hypophosphites was owing to their *impurity*, I only expressed the conclusion to which I had been led by experimental investigation."

"I had found by repeated clinical experiments that when a

certain proportion of alkaline carbonates was added to the pure hypophosphites, the physiological and therapeutical effects were manifested but *slowly* and *incompletely*, or *failed altogether*."—*Churchill*, p. 56.

"I found the commercial hypophosphite of lime, as sold by the trade, to contain phosphite, phosphate, or carbonate; that it is occasionally adulterated with chloride of sodium, sulphate of lime, carbonate of magnesia, and oxide of zinc."—*Janssen Répertoire de Chimie.*

"We have frequently met with *adulterated samples* of hypophosphite of soda in commerce. Some contained carbonate of soda, others free soda. In some instances we have even found sulphuret of sodium and carbonate of lime.

"One sample contained so much sulphuret, indeed, that they looked quite yellow and gave out a decided smell of hydrosulphuric acid (*rotten eggs*), when exposed to the atmosphere." *Heyward's Journal of Medicine*, p. 85.

"All physicians who have employed the hypophosphites with success have pointed out the absolute necessity of employing them perfectly *pure.* In almost every instance in which we have examined specimens of the hypophosphites of lime and soda as usually employed, we have found these salts *to be impure.* The hypophosphite of soda has been alkaline in reaction, containing sometimes free alkali, not unfrequently carbonate of soda. The salt of lime has contained smaller or larger quantities of free lime, phosphate of lime, and occasionally carbonate of lime."—*London Lancet.*

Dr. Parigot, Professor at the University of Brussels, observes: "When the hypophosphites were first introduced by Dr. Churchill, all the samples were equally good and active, wherever procured, because at that time there was no great demand for it; but since it has come into very general use, manufacturers have become less scrupulous in preparing it."

Dr. Maestre, Professor of Clinical Medicine in the University of Granada, Spain, sums up the result of his experience in the following words: "1. The hypophosphite of soda and lime must be *perfectly pure;* this is one reason of the advantageous results obtained by me."

From what precedes, therefore, and much more might be added, it is evident that the method of manufacturing, upon which the chemical purity depends, is of the greatest importance; and chemical analysis will also show that no reliance can be placed upon the hypophosphites, as therapeutical agents, if kept for any length of time in the state of *salts*, and *still less if in solution.* In my preparation, the hypophosphites are in the form of a syrup, made from beet sugar, perfectly free from lime, and by cold maceration, and *this is the only reliable form in which they can be kept free from decomposition.*

I venture to say that the most of the so-called hypophosphites now in the market are not *true hypophosphites;* that if they do contain any of the phosphatic element, it is the electro-negative principle of the phos*phates*, and not that of the hypophos*phites*. They have no affinity for oxygen, and are, therefore, *negative* in the treatment of consumption.

My Syrup of the Hypophosphites of LIME AND SODA is CHEMICALLY PURE, and Dr. Churchill distinctly says "that in the treatment of phthisis these hypophosphites alone should be used." My Syrup is not a *secret*, or copyrighted nostrum. My formula is given in full, and I claim the advantage over other makers in the skill which years of experience can alone impart. I therefore respectfully ask the medical profession to give my Syrup a fair trial in their practice, believing they will not be disappointed in the results. For the formula by which my Hypophosphites are prepared I refer to Professors Stohmann and Engler's German edition of *Payen's Precis de Chimie Industrielle*, and also to Dr. Churchill's work.

Dr. Williams says, "The introduction of *these hypophosphites* into the blood produces a glowing influence — as a respiratory excitant, expanding the chest; as a pyrogenic, increasing animal heat and nerve force, and removing erratic pains; and as a hæmatogen, forming a nucleus for the rallying of red globules; it increases appetite and cheerfulness, and controls expectoration, night sweats, and diarrhœa.

COD-LIVER OIL EMULSIONS.

At least one-half of the failures in the use of the hypophosphites are owing to no other cause than their administration in combination with cod-liver oil; that is, in the form of any of the numerous emulsions, for these preparations are merely saponified oils, and are not assimilated without great difficulty, the natural condition of the oil being changed to a liquid soap.—*Dr. Churchill*, p. 141.

Dr. Mayhofer says: "Cod-liver oil justly merits the high reputation which it has acquired in correcting these deficiencies of nutrition, commonly comprehended as scrofulosis and tuberculosis, *but it should never be given in the form of emulsion with the hypophosphites.*"

Dr. Williams, in his late work on Pulmonary Consumption, says with cod-liver oil alone he has not cured two per cent. of the cases in which it had been used; and continues, "It has happened to me on several occasions that a patient has long been taking cod-liver oil, and, after having derived benefit from it, halts in his improvement, or even loses ground, and then the hypophosphites have been followed by a marked change for the better."

Alcohol is often employed, for the same purpose as cod-liver oil, and with similar results—that of merely retarding the progress of the disease.

There can be no doubt, however, as to the salutary influence exercised by cod-liver oil in patients exhibiting a strumous diathesis, of a slender and lean figure, and thin, transparent skin, with weak pulse and great excitability of the nervous system; but it should never be used as an *emulsion* with *alkaline hypophosphites.* When it is desirable to give cod-liver oil *with the hypophosphites*—not in the form of emulsion—the pure cod-liver oil should be exhibited clear, or floated on claret wine, porter, lemon, or orange juice, or in any way the physician may direct. The following combination is suggested. It is pleasant to take and easily digested:

℞ Cod Liver Oil, 8 ounces.
Whiskey or Jamaica Rum, 8 ounces.
Juice of 3 Lemons.
Ess. Bitter Almond, 40 drops.
Sugar if desired.

DOSE:—One small wineglassful immediately after breakfast and supper, with full doses of the Syrup Hypophosphites one hour later.

When alcohol is indicated, the following combination is a good one:

℞ Whiskey, Rye or Scotch, 8 ounces.
McArthur's Syrup, one bottle.

Mix. Dose:—Three to four tablespoonfuls in the course of the day in divided doses, diluted with cold or hot water. Lemon may be added.

CHILDREN.

"**The Hypophosphites Prevent and Cure Consumption in Children.**—I have had constant opportunity of observing the action of the hypophosphites of *lime* and *soda* in delicate children, especially those born of consumptive parents, and I feel bound to state that I have never seen *one* single instance of such children becoming consumptive, when the remedy has been used in a proper manner. The hypophosphites also exert a *special action* in promoting the *growth* of children. I have had a large number of children under my care, and may state broadly, that recovery in children under adult age, at all stages of consumption, whenever the amount of sound lung tissue remains sufficient to support life, takes place in nine cases out of ten. I might quote instances of children, where the disease had already reached the third stage, who have recovered under the hypophosphites, and have been in the enjoyment of sound health for many years.

"**Teething Children.**—**Convulsions.**—In the first teething of children the hypophosphites produce an heroic effect, and, if properly used, will act as a preservative against the accidents of this difficult period of life. When given to teething children who are pale, peevish, sad, emaciated, without appetite or strength, suffering from fever and diarrhœa, loss of sleep, and apparently in imminent danger of convulsions, I have never seen a single case, where the whole of these symptoms have not yielded to a few doses of Syrup of Hypophosphites of Lime and Soda, and the evolution of the teeth afterward proceed as if in perfect health."—*Churchill*, p. 80.

Results in my practice verify the above, and particularly in the case of my own little girl. The dose should seldom exceed one small teaspoonful in twenty-four hours, and be so divided as to give one-quarter every sixth hour, mixed with a little cold water.

"**In tabes Mesenterica**, Dr. Purdon has found that the hypophosphites act slowly but surely. He considers that they act by dissolving the tuberculous matter deposited in the folds of the mesentery and mesenteric glands; possibly by causing disintegration of the fibrine. In the *remittent fevers* of childhood Dr. Purdon employed the hypophosphites with marked success. They seem to fulfil, he says, all the required indications, in causing sleep, relieving thirst, cleansing the tongue, increasing the appetite, and arresting any resisting intestinal disorder, in a much

shorter time than can be obtained by any other remedy."
—*Braithwaite's Retrospect.*

For children cutting their teeth, the Syrup given in doses of one-half to one small teaspoonful in half a wineglassful of water, in the twenty-four hours, and be so divided as to give one-quarter every sixth hour, will be found to produce immediately a beneficial effect.

"It gives vigor and strength to nursing mothers, imparting the necessary elements to their milk, for the production of bone and sinew in the child, besides preventing in many cases that draining of the system brought about by improper feeding, badly ventilated dwellings, and excessive lactation."—*British Medical Journal,* 1880.

The writer has found these hypophosphites highly useful in complaints of infancy connected with the scrofulous diathesis and defect in the osseous system.

PRECAUTIONS TO BE ADOPTED IN THE TREATMENT OF CONSUMPTION.

At the International Congress of Hygiene held at La Have, in September, the following conclusions were adopted regarding the treatment of phthisical patients:—

"It is demonstrated that pulmonary phthisis can be, in certain cases, transmitted from the sick to individuals in health. Although the chances of this transmission are limited, prudence requires certain precautions.

"1.—No one should be allowed to share the sleeping-chamber or the bed of a tubercular patient in an advanced stage of the disease. The apartment of a phthisical individual should be constantly aired and ventilated.

"2—The danger resides especially in the sputa, which should not be allowed to go on the floor or on clothing, where they may dry and become converted into dust.

"3—The sleeping-rooms, the bed-linen, and clothing which have been used by consumptives, should always be disinfected. Steam (at 100° C.) and washing in boiling water are the best means of disinfection.

"4—Convalescents from chest disorders and feeble and exhausted patients should especially avoid prolonged contact with the tuberculous."—*Revue de Therapeutique Med-Chir.*, No. 23.

For the administration of phosphorus there is not a more certain, a more efficient or safer medium than the Chemically Pure Hypophosphites, and the adaptation of DR. MCARTHUR'S SYRUP to the treatment of *Nervous Affections* renders them very useful as *adjuncts* in a great variety of diseases, inasmuch as the *Hypophosphites* are more readily absorbed by the system than all other preparations of *phosphorus.*

DR. DONALD MCGREGOR writes as follows:

"In the '*Consequences*' *of Spermatorrhœa*, where there is general weakness and nervous irritability, DR. MCAR-

THUR'S SYRUP IS A CAPITAL REMEDY, and in *impotence*, where there is want of sexual vigor, etc., in *old* or *young*, I have never had anything serve me so well as the following:

℞ Syr: Hypophos: Comp: McArthur, one bottle.
Fld. Ext. Damiana, 2 ounces.

M. Sig.: Two teaspoonfuls three or four times a day.

The writer has found the following prescription singularly useful in spermatorrhœa and *impotence:*

℞ Liquor Strychniæ, ʒij.
Fld, Ext. Coca, ℥j.
Syr: Hypophos: Comp: McArthur, one bottle.

M. Sig. Dose:—Two to three teaspoonfuls before each meal, either clear or in a little port wine or water.

When *strychniæ* or *nux vomica* is indicated, the following makes an elegant preparation, and will be found excellent in dyspepsia, with headache, lassitude, and constipation. It has also been employed with *asserted* success in neuralgia and chorea, and will promptly restore to its normal beat the arhythmic pulse so frequently found in aged patients.

℞ Liquor Strychniæ=ʒij. (equal to one grain of Strychniæ.)
Syr: Hypophos: Comp: McArthur, one bottle.

M. Sig. Dose.—Two teaspoonfuls three times daily to adults.

NOTE.—The effects of Strychniæ upon the system are identical in character with those of Nux Vomica. Under the head of TINCTURA NUCIS VOMICÆ, the "Dispensatory of the United States" says: "The tincture is not an eligible form for administering Nux Vomica, as it is equally uncertain with the medicine in substance, and has the disadvantage of excessive bitterness. Strychniæ is preferable."

Physicians will find the following a most agreeable and effective mode of administering a pure bitter tonic, with the Syrup Hypophosphites, when such is indicated. It excites the appetite, invigorates digestion, and agrees with the most delicate stomachs.

℞ Syr: Hypophos: Comp: McArthur, one bottle.
Tinct. Gentian, ℥ij.

M. Sig. Dose for adult, a dessertspoonful three times a day, before meals, in a little water.

In cases where a powerful and prompt *ferruginous tonic* is required, I am in the habit of writing thus:—

℞ Ferri et Quiniæ Citras, ʒj.
Syr: Hypophos: Comp: McArthur, one bottle.

M. Sig. Dose.—Two teaspoonfuls in water, to adults, during the meal. For children, according to age.

The above prescription is particularly recommended in cases of anæmia, or poorness of the blood; in general debility and loss of strength; in the convalescence of some acute diseases, when all febrile symptoms have disappeared; in the relaxed and debilitated state of the system occasioned by *uterine affections*, such as chronic *leucorrhœa* and *spermatorrhœa;* and it may be relied upon for the treatment of irregularity in the circulation of the blood in young females, accompanied by paleness of the complexion, etc. It increases the red globules of the blood, and powerfully aids in the formation of hæmoglobin.

These *Hypophosphites* are antagonistic to intemperance and to the opium habit, and the use of them will do away with that feeling of weakness and depression that impels to the use of alcoholic beverages and narcotics, the following being an excellent and *proved* prescription:—

℞ Syr: Hypophos: Comp: McArthur, one bottle.
Tinctura Cinchonæ, ℥ij. to ℥iv. Vel. Tinct. Coca.

M. S. Dose.—A tablespoonful in 1-2 tumbler of cold water three times daily, either before, after, or with the meal.

The following combination affords one of the most efficient and elegant tonics to be found in the whole range of the *Pharmacopœia*. It not only improves the appetite, but stimulates digestion and assimilation.

℞ Ferri et Strychniæ Citras, ʒj.
Syr: Hypophos: Comp: McArthur, one bottle.

Sig. Dose.—Two teaspoonfuls in water, to adults, before, after or with meals.

CODEIA AND THE SYR. HYPOPHOSPHITES *as a Sedative in advanced Phthisis.*

℞ Codeia, grs. xvi.
Syr: Hypophos: McArthur, one bottle.

The DOSE is about two teaspoonfuls three times a day, to be increased or diminished as the case requires. It allays cough without disturbing the digestive system, and is tolerated when opium and its other alkaloids are not. In many troublesome coughs, particularly if depending on catarrh of the trachea or bronchi, I am in the habit of prescribing the above mixture with all the *benefit* and none of the *ill* effects of opium.

Prof. JACKSON strongly recommends the use of the following mixture in coughs and bronchial affections:

℞ Fluid Ext. Pruni Virg., one ounce.
Rum, Whiskey or Cognac, one ounce.
Syr: Hypophos: Comp: McArthur, one bottle.

M. S. Dose:—Two tablespoonfuls in the course of the day in divided doses.

"The mixture is most agreeable to the taste and well tolerated by the stomach."—*Medical Journal.*

The following has just been communicated to me by one of our leading Lynn physicians:

℞ Syr: Hypophos: Comp: McArthur, one bottle.
Best Old Whiskey, eight ounces.

DOSE.—One tablespoonful three or four times a day, diluted with hot or cold water. He says, "with this medicine I know I have arrested and probably cured *four cases of incipient consumption* during the last year."

Syr: Hypophos: Comp: McArthur.

CHEMICALLY PURE

GENERAL DIRECTIONS FOR USE,—DOSE, ETC.

There is no fixed or invariable dose. My *Syrup of the Hypophosphites of Lime and Soda* combines with a neutral syrup twelve and one-half centigrammes of the hypophosphites to each teaspoonful.

As a general rule, for an adult, begin with two teaspoonfuls morning, noon and night, before, after, or with the meal, and increase gradually to a tablespoonful, *three times daily.*

In case of females of very delicate constitutions, leading a sedentary life and not used to much physical exertion, the above dose should be reduced from one-quarter to one-half.

For children from eight to thirteen, the dose is the same as for delicate females.

For a child from two to seven years of age, the dose should be from *one-third* to one teaspoonful.

For *teething* infants the dose should seldom exceed one

small teaspoonful in twenty-four hours, and be so divided as to give one-quarter every sixth hour. The dose should be regulated, however, by the careful advice of the physician. The Syrup, if given as above to children cutting their teeth, will be found to produce a most beneficial effect.

The Syrup is free from any medicinal taste whatever, and may be taken alone, or mixed with any of the patient's usual beverages, such as milk, tea, coffee, cold water, etc.

In all cases where alcoholic stimulants are indicated, any kind of pure spirituous liquors (except acid wines) may be added to suit the taste and requirements of each case.

The Syrup should be taken three times a day, before, after, or with the meal.

A very agreeable and refreshing drink may be made by adding lemon juice to the Syrup and water.

NOTE.—Many physicians recommend this method to their patients, particularly those to whom sweet is an objection. The lemon is *not* incompatible with the action of the Hypophosphites.

It will often be found advisable, after the patient has taken the medicine a week or more, to omit the dose for a day, and see how he fares without it; then resume, either in the same or in a smaller or larger dose, according to the indications presented.

Of course the medical attendant will increase or diminish the above doses according to circumstances.

Physicians when prescribing will please write thus:

℞ **Syr: Hypophos: Comp: McArthur. One bottle.**

SUGGESTIONS.

It is not reasonable to expect that a remedy, which acts through the functions of nutrition by promoting the building up of new and healthy tissues, and by eliminating those which are diseased, should, in all instances, manifest its action in a few days.

When the lungs are only slightly affected, or when the disease is of recent origin, this early and immediate action will often be met with. Such, for instance, is the case in acute phthisis or catarrhal pneumonia. Otherwise, when the local disease is extensive, or has already lasted some time, the action of the remedy must be chronic like the disease itself.

As there is very great variability in the immediate effect produced upon different patients, the physician should therefore carefully feel his way, increasing the dose every second or third day until some apparent effect is produced either upon one, several, or the whole of the symptoms. When once this has been obtained, and the treatment has been continued for about a week, it will often be found advisable to omit the treatment for one or two days, and see how the patient fares without it. After this it should be resumed, either in the same or in smaller or larger dose, according to the indications presented.

The interval may be gradually increased with improvement in the patient's health, and his showing signs of physiogenic plethora. When all the general symptoms (weakness, emaciation, etc.) have disappeared, and nothing remains but those depending upon the local condition, such as cough expectoration, etc., two or three doses, sometimes even only one dose a week will be found sufficient to keep the patient in the state of physiogenic plethora necessary for the completion of the cure.

But, as already said, this will vary with every individual case. Some patients require a daily dose of three or four teaspoonfuls and cannot do with less; others feel better with even one teaspoonful.

When the cure is once complete, when the local lesions have disappeared, or have cicatrized, the patient should continue to take one or two doses a week as a prophylactic. Many patients find that they cannot leave off the treatment altogether for a longer time than three or four weeks without feeling the want of it, and getting below *par* particularly if they remain in the same hygienic conditions (such as over-work, etc.) as those which originated the complaint. I have even met with some who could not leave it off for a single day without feeling the want of it.

Dr. Barella, in an article in "Le Scalpel," December, states the current opinion on the continent with regard to the hypophosphites: "There is not a single organ of the French or foreign medical press, which has not published cases *conclusive* as to the therapeutical value of the hypophosphites *in every stage of* consumption. Many of these cases have been collected in the public hospitals. In the present state of science, and although we have to deal with a disease which has been looked upon as incurable, and which carries off one-third of the adult population, *it is a fact, which nobody will now deny, that the treatment by the hypophosphites produces remarkable and, in some cases, immediate effect.* The cases of cure are the more numerous in proportion as the disease itself is less advanced. Discussion has been exhausted as to Dr. Churchill's doctrine, which, as we know, is founded on the principle of stœchiology, that is to say, upon the study of the constituent elements of the proximate principles of which our system is built up. Dr. Churchill *does not pretend, as has been asserted, to cure those who are already dying.* His treatment (by the hypophosphites) is the specific remedy for the diathesis or general condition, and *not for the local disorganization* to which it leads. *He therefore pretends to cure only under certain conditions,* and by following certain rules, which he has minutely explained in his work on the subject.

McARTHUR'S

SYRUP HYPOPHOSPHITES COMP: C. P.

is kept in stock by the wholesale druggists throughout the United States, to whom all orders for less than gross lots should be sent.

TO PHYSICIANS.

If your retail druggist does not have McARTHUR'S SYRUP HYPOPHOSPHITES COMP: C. P. in stock, have him order it for you, or order it yourself from your nearest wholesale druggist.

Any physician desiring to test McARTHUR'S SYRUP HYPOPHOSPHITES COMPOUND, C. P., will be furnished a sample bottle without expense, except express charges.

NOTE.—We cannot guarantee the genuineness of our Hypophosphites except when purchased in original bottles. Then, if desired, the trade label may be removed and prescription-directions substituted in its place.

As it is made ***only*** *for physicians, there are no printed wrappers or advertisements about the bottle.*

We shall esteem it a favor if any member of the profession should desire to correspond with us upon the subject of the "*Hypophosphites.*" Such correspondence to be used only in accordance with the Code of medical ethics.

Office Address:

McARTHUR HYPOPHOSPHITE CO.,

BOSTON, MASS.

Personal Address:

J. A. McARTHUR, M. D.,

31 SOUTH COMMON STREET,

LYNN, MASS.

General Agents.—POTTER DRUG AND CHEMICAL CO., 135 & 137 Columbus Ave., Boston; F. NEWBERY & SONS, 1 King Edward St., London, E. C., England; NORTHROP & LYMAN CO., 40 Scott St., Toronto, Canada; R. TOWNS & CO., Sydney, New South Wales; FELTON, GRIMWADE & CO., Melbourne, Australia.

McARTHUR'S
SYRUP
OF THE
HYPOPHOSPHITES
COMP. (LIME & SODA)
CHEMICALLY PURE
FORMULA OF
DR. CHURCHILL
OF PARIS.
PREPARED
M ARTHUR
HYPOPHOSPHITE
BOSTON

www.ingramcontent.com/pod-product-compliance
Lightning Source LLC
Chambersburg PA
CBHW022012190326
41519CB00010B/1491
* 9 7 8 3 3 3 7 8 1 7 0 2 2 *